LEARN ABOUT
PYRAMIDS

PETER MELLETT

MY FATHER'S WORLD®

This edition published in 2019 by
My Father's World,
P.O. Box 2140,
Rolla, MO, 65402, U.S.A
www.mfwbooks.com

Produced by Anness Publishing Ltd
Algores Way, Wisbech,
Cambs, PE13 2TQ, England
info@anness.com
www.annesspublishing.com

Anness Publishing has a new picture agency outlet for images for
publishing, promotions or advertising. Please visit our website
www.practicalpictures.com for more information.

Publisher: Joanna Lorenz
Managing Editor, Children's Books: Sue Grabham
Editor: Ann Kay
Designer: Caroline Grimshaw
Consultants: Dr Anne Millard, Jack Challoner
Photographer: John Freeman
Stylist: Melanie Williams
Picture Researcher: Annabel Ossel
Illustrators: Peter Bull Art Studio, Stuart Carter, Simon Gurr,
Stephen Sweet/Simon Girling and Associates

Manufacturer: Anness Publishing Ltd,
Algores Way, Wisbech, Cambs, PE 2TQ, England
For Product Tracking go to:
www.annesspublishing.com/tracking
Batch: 0880-24752-1104

The publisher would like to thank the following children, and their
parents, for modeling in this book: Steve Jason Aristizabal, Ricky Edward
Garrett, Sophie Halliström, Mitzi Johanna Hooper, Imran Miah,
Jessica Moxley, Kim Peterson, Marlon Stewart. Special thanks also to
Bob Partridge for all his patient help.

PUBLISHER'S NOTE
Although the advice and information in this book are believed to be
accurate and true at the time of going to press, neither the authors
nor the publisher can accept any legal responsibility or liability for
any errors or omissions that may have been made nor for any
inaccuracies nor for any loss, harm or injury that comes about from
following instructions or advice in this book.

PYRAMIDS

CONTENTS

WHAT ARE PYRAMIDS?

S HUT your eyes and say the word *pyramid* to yourself several times. Exactly what picture comes into your mind? Perhaps you start to think about the huge stone shapes and buildings that stand by the great river Nile in Egypt or stand in parts of Central America. It may be that you remember seeing some kind of pyramid shape at the top of a modern skyscraper. Perhaps you have looked through a microscope at tiny, pyramid-shaped crystals in a science class at school. As you can see, there are many, many different kinds of pyramids all around us. Some pyramids are made by people and others are natural, but they all have one thing in common— the same special shape. All pyramids stand on a flat base and have flat, sloping sides that are called faces. These sloping faces are always triangular in shape, and they meet at one point at the top of the pyramid. So, when you hear the word *pyramid*, the first thing to think about is this special shape.

The most common kind of pyramid has a square, flat base and four sloping, triangular sides. Most of the pyramid-shaped buildings constructed during the last 5,000 years have this shape.

The exterior was once covered in limestone, which is now missing

The Giza pyramids
These pyramids have stood by the Nile in northern Egypt for 4,600 years. They were built as tombs for great kings and queens. The exteriors were once covered in a layer of smooth stone, but robbers stole the stones.

Temple II, Tikal, Guatemala

This pyramid is in Central America, on the other side of the world from Egypt. It has steps up the outside and a temple at the top. Most Central American pyramids were built between 600 and 1,500 years ago.

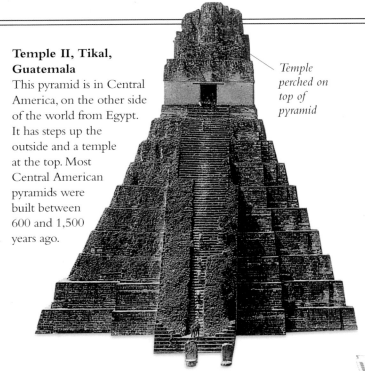

Temple perched on top of pyramid

Canary Wharf, London

This tower was built during the 1980s and has a pyramid at the top. It forms a striking landmark that can be seen for miles. The pyramid shape is simple and dramatic. This means that it catches your attention. It has been used by many modern architects.

Spinel crystal

Not all pyramids are made by people. There are also pyramids in the natural world. This crystal is shaped like a pyramid. It is made of a mineral called spinel and is formed over many years, deep underground. Natural crystals also occur in other shapes. Salt crystals, for example, are cubes.

FACT BOX

- Egyptian pyramids were built from millions of stone blocks, or mud bricks encased in stone.

- Most Central American pyramids have a core of rubble, with stone blocks on the outside.

- Egyptian pyramids were built as tombs. Most Central American ones were used as temples.

PYRAMIDS AND MATH

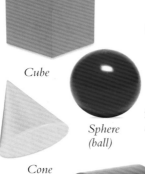

Cube

Sphere (ball)

Cone

Cylinder

There are many different geometrical shapes. Try comparing them with a pyramid.

THE kind of math that involves studying shapes is called geometry. Solid geometrical shapes include cubes, cones, spheres, cylinders—and pyramids. There is a whole family of different shapes that we can call pyramids. The simplest members of this family are pyramids with just four faces. They stand on a triangular base and have three other triangular faces. These pyramids are called tetrahedrons. The next members of the family have a total of five faces—they stand on a square base and have four triangular faces. These pyramids are called square pyramids. You can make other pyramid shapes by adding more sides to the base and more triangular faces. You could make a pyramid with hundreds of faces, but most natural pyramids and pyramids built by people are either tetrahedrons or square-based. Every pyramid has two important measurements that describe its shape. The base length measures along one side of the base. The vertical height measures straight upward—from the middle of the base to the point at the top.

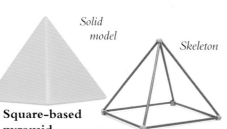

Solid model

Skeleton

Square-based pyramid
Here you can see a solid and a skeleton model of a square-based pyramid. This type of pyramid stands on a square base. It is also called a regular pyramid because all the triangular faces are the same shape and size. It has a total of eight edges.

Tetrahedron (Triangular-based pyramid)
A solid and a skeleton model of a triangular-based pyramid. A tetrahedron is a pyramid that stands on a triangular base and has three other triangular faces. It has four faces and six edges.

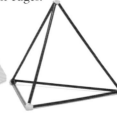

Short and tall
Shapes like this are also pyramids, although you might not think they are when you first look at them. See how many pyramid shapes you can spot around you.

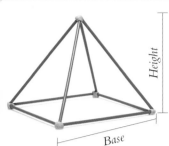

Square-based pyramid

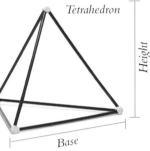

Tetrahedron

The base length measures along one side of the base. The height measures from the top straight down to the base.

Height and base

Make a square pyramid and a tetrahedron from straws and reusable adhesive. Now try measuring them to learn about their shapes. It is easier to measure their height if you place the ruler in the center of the base.

The volume of a cube is base length x base length x height.

Put the pyramid inside the cube and see how it takes up one third of the cube.

Volumes of pyramids

The volume of a shape is the amount of space inside it. Make a straw cube. Now make a square-based pyramid with a base-size and height the same as your cube. The volume of the cube should be three times the volume of the pyramid.

The volume of a square pyramid is ⅓ of base length x base length x height.

Tessellation

Some pyramid shapes tessellate (fit together without any gaps). Experiment with small cardboard pyramids you have made yourself or with the pyramid shapes from egg cartons.

SHAPE AND STRENGTH

W HY are pyramid shapes special? Why did ancient civilizations build gigantic pyramids rather than massive cubes or rectangles? These projects will help you find out. In the first project, change a cube into a pyramid that is three times taller than the cube. The second project shows you that pyramids are much more rigid than cubes and do not collapse so easily. Seven hundred years ago, an earthquake destroyed the Egyptian city of Cairo, but the nearby pyramids were hardly affected. So, if you want to build a huge, impressive shape with the smallest amount of material, choose a pyramid. Many centuries later, by the 1800s, modern building materials were developed and people could build tall, rectangular buildings that were strong and stable.

You will need: modeling clay, ruler, plastic modeling knife.

Try making your own pyramid out of cardboard. Copy this shape onto cardboard. Cut it out, fold along the dotted lines and glue the tabs.

Bases and heights

1 Make two cubes from modeling clay. The faces should measure about 1 1/2–2 inches, but they must all be the same. Use a ruler to check your measurements.

2 Reshape one of the cubes to form a tall, square-based pyramid. Its base must be the same size as the original cube. You now have a cube and a pyramid.

3 Now measure the cube and pyramid. They have the same volume and the same size base, but the pyramid is three times taller.

8

A question of strength

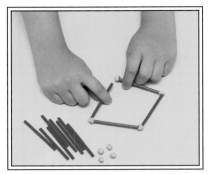

1 For this project, you will need two models—a cube and a square pyramid. Make them out of large plastic drinking straws and reusable adhesive. First, make the cube.

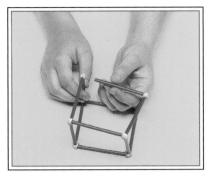

2 After attaching four straws to make the base of your cube, you will need eight more to finish it. Make sure that your cube is even, with each face the same size.

3 Now make a square pyramid. The base should be the same size as the base of your cube. Make the base and attach four more straws to it to complete your model pyramid.

M A T E R I A L S

You will need: plastic drinking straws (cut in half), reusable adhesive.

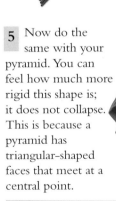

4 Push down gently with your hand over the center of the cube. Move your hand slightly to one side as you push and feel the cube start to collapse.

5 Now do the same with your pyramid. You can feel how much more rigid this shape is; it does not collapse. This is because a pyramid has triangular-shaped faces that meet at a central point.

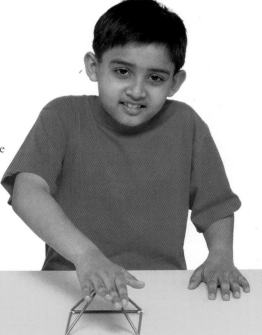

THE AGE OF PYRAMIDS

For more than 4,500 years, people built huge pyramids all over the world. The oldest Egyptian pyramids were started almost 4,600 years ago. The youngest pyramids in Central America were finished about 600 years ago. All these pyramids were connected in some way to the religion of the people who built them. Ancient Egyptians buried dead kings and queens inside pyramids. The Aztec, Toltec and Mayan people of Central America worshipped their gods in temples at the tops of pyramids. In Burma and Java there are huge, pyramid-shaped temples that people still use for worship today. Building these pyramids took a long time—sometimes over 50 years. It often involved thousands of people. We can rarely afford to tackle such huge projects today.

Santa Cecilia, Mexico
The Santa Cecilia pyramid in Mexico has a flat top with a temple on it. This style is common in Central America.

Earth mound, North America
The Hopewell people of North America built huge, conical mounds of earth. Some were 325 feet wide. Temples or palaces stood on the tops of the mounds. Today, only grass-covered hillocks remain.

Temple of the Sun, Peru
This pyramid-shaped mound is called the Temple of the Sun. It was built by the Moche people in northern Peru nearly 2,000 years ago. It contains a massive 143 million adobe bricks—bricks made from sun-dried mud.

		3000		2500		2000		1500	
AFRICA		Archaic Period	[Giza]	Old Kingdom	EGYPT	Middle Kingdom		New Kingdom	
NORTH & SOUTH AMERICA									Olmecs
ASIA		Sumer				Babylonians and Assyrians		MIDDLE EAST	

Hindu temple, Java
There are many pyramid-shaped buildings in the Far East. This Hindu temple, at Prambanan in Java, is over 1,000 years old. It has several pyramid-shaped roofs. These are made from stone that has been carved into beautiful patterns.

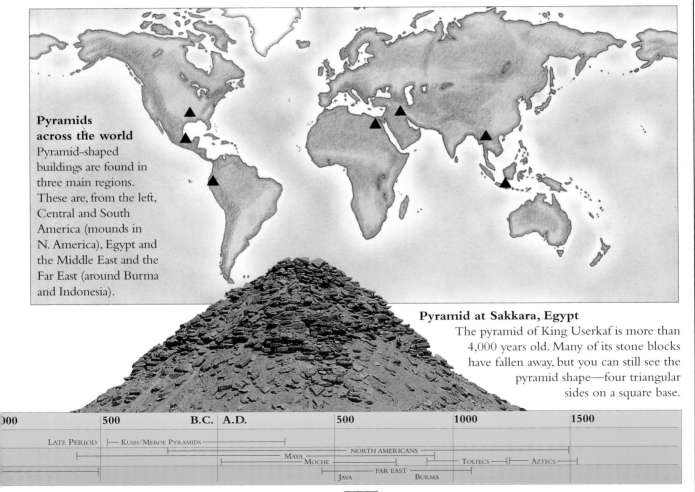

Pyramids across the world
Pyramid-shaped buildings are found in three main regions. These are, from the left, Central and South America (mounds in N. America), Egypt and the Middle East and the Far East (around Burma and Indonesia).

Pyramid at Sakkara, Egypt
The pyramid of King Userkaf is more than 4,000 years old. Many of its stone blocks have fallen away, but you can still see the pyramid shape—four triangular sides on a square base.

000	500	B.C.	A.D.	500	1000	1500

LATE PERIOD — KUSH/MEROE PYRAMIDS —

NORTH AMERICANS

MAYA

MOCHE

TOLTECS — AZTECS —

FAR EAST

JAVA BURMA

UNCOVERING THE PAST

Modern archaeologists sieve the earth and look through it for clues using delicate instruments such as tweezers. They dig the ground with tiny trowels and move sand away with soft brushes.

ARCHAEOLOGISTS are people who try to understand the past by searching for ancient clues. Today, they work very carefully, using special scientific instruments so they do not damage evidence. They usually need permission from the government of the country they are working in before they are able to investigate a site. However, this was not always the case. As recently as 100 years ago, archaeologists caused a lot of damage. They took away precious finds to fill museums and make themselves famous. They were not much better than tomb-robbers, who have also caused a great deal of damage. For thousands of years, robbers have assumed that, because the pyramids were built by rich, powerful rulers, that fabulous treasures must lie hidden inside. They have broken into pyramids and taken away anything they found. Some hacked into the stone looking for hidden tunnels. In more recent times, they even used explosives. These robbers have destroyed important clues about ancient life. This is why methods have to be so painstaking today.

Entering by force
This shows the terrible damage done by early archaeologists to Khafre's pyramid at Giza, in Egypt. King Khafre built his pyramid with a false door, designed to mislead robbers. Archaeologists trying to find a way in used pickaxs and even gunpowder to enlarge the entrance. This picture was painted in 1822, before photography was invented.

The Grand Gallery

There is a steep passage inside Khufu's pyramid at Giza. This passage leads to the burial chamber and is called the Grand Gallery. After Khufu's funeral, the passage was sealed with huge stone blocks. Just a few years later, Ancient Egyptian tomb-robbers broke past the blocks and stole everything.

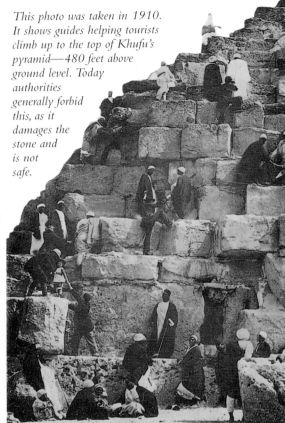

This photo was taken in 1910. It shows guides helping tourists climb up to the top of Khufu's pyramid—480 feet above ground level. Today authorities generally forbid this, as it damages the stone and is not safe.

X-ray

Modern methods

Today, X-ray and scanning machines can look right inside objects such as ancient Egyptian mummies. Ancient Egyptians preserved the bodies of their rulers by drying them and wrapping them in bandages. These machines can see the skeleton beneath the wrappings.

Scanning machine

FACT BOX

• Today, tiny robots can be used to take TV cameras into small spaces. For example, a robot explored shafts inside Khufu's pyramid.

• Discoveries can be made by beaming radio waves through soil and rock. The waves bounce back to the surface and reveal hidden chambers and passages.

ANCIENT EGYPT

The most impressive Egyptian pyramids that we see today were built in the northern part of the region.

THE ancient Egyptian civilization began about 5,000 years ago and lasted for 3,000 years. Its people lived on the fertile lands around the Nile river. Beyond this area lay parched desert, which protected the country from warring neighbors. Every year, the water level of the river rose and flooded over the banks. This flood-water brought huge amounts of fresh soil. Farmers grew plentiful harvests of food in the rich soil. Compared to other places in the world, ancient Egypt was a very good place to live. The people believed that they were especially well looked after by the gods and that their king (also called a pharaoh) was a god living on earth. When a king died, the priests said he must return to the sky and rejoin the gods. The king's pyramid was a very special tomb. It was believed to be the place where his spirit left the earth and started its journey into the heavens.

Life along the Nile
This tomb-painting shows the Underworld—where the Egyptians believed people went after death. The underworld was an idealized version of life in ancient Egypt, and gives us an idea of what life was like on the shores of the Nile. You can see palm trees along the river and irrigation channels bringing water to the fields. At the top, on the left, the deceased person and his wife worship various gods.

KING		DJOSER	SEKHEMKHET	HUNI		SNEFERU	KHUFU	KHAFRE	MENKAWRE	USERKAF
PYRAMID ▲ = PYRAMID		▲ STEP PYRAMID	▲ INCOMPLETE PYRAMID	▲ COLLAPSED PYRAMID		▲ BENT PYRAMID	⊢——— ▲ GIZA COMPLEX ———⊣			
DATES B.C.		2700 START OF THE AGE OF THE GREAT PYRAMIDS		2650		2600		2550		2500

←——— 3000 B.C.: START OF THE ANCIENT EGYPTIAN CIVILIZATION.

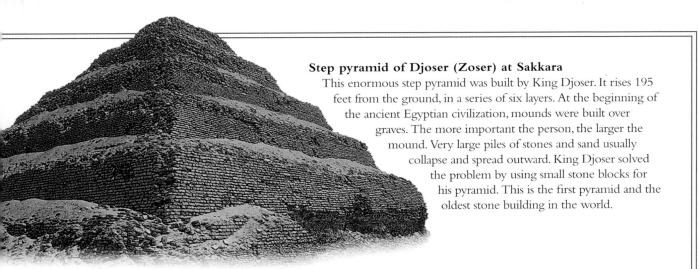

Step pyramid of Djoser (Zoser) at Sakkara

This enormous step pyramid was built by King Djoser. It rises 195 feet from the ground, in a series of six layers. At the beginning of the ancient Egyptian civilization, mounds were built over graves. The more important the person, the larger the mound. Very large piles of stones and sand usually collapse and spread outward. King Djoser solved the problem by using small stone blocks for his pyramid. This is the first pyramid and the oldest stone building in the world.

Collapsed pyramid of Huni at Meidum

This step pyramid was built by King Huni. Notice how steep the sides are. His son, Sneferu, added angled stones to the outside to make the first true, smooth-sided pyramid. This casing crashed down, unfortunately, bringing much of Huni's pyramid with it.

Sneferu's bent pyramid at Dashur

This pyramid was built by Sneferu. When it was half-done, the builders realized that the sides were too steep. To prevent it from collapsing, they made the top slope more gently.

SAHURE	NYUSERRE	UNAS		PYRAMIDS MOSTLY BECOME MUCH SMALLER IN SCALE. →
POOR QUALITY BUILDING—ONLY 4 OUT OF 14 PYRAMIDS SURVIVE		PYRAMID TEXTS		
2450	2400	2350	2300	100 BC: END OF ANCIENT EGYPTIAN CIVILIZATION. →

THE PYRAMIDS OF GIZA

GIZA, in northern Egypt, has some of the finest pyramids in the world. There are three kings' pyramids here—built by King Khufu, King Khafre and King Menkawre. The largest, King Khufu's, is the largest pyramid ever built in Egypt—originally 480 feet high with a base 760 feet long. Khufu became king in 2589 B.C. He ruled for 23 years, and his reign came near the start of the Ancient Egyptian civilization. Like all kings at that time, the first thing he did was to decide where to build his pyramid tomb. Each king had to make sure his spirit could return to the sky when he died, and his pathway to heaven was his pyramid. The people believed that the deceased king continued to care for them from heaven. The Nile would continue to flood each year and good harvests would grow. The building of Khufu's pyramid, which lasted for all of his reign, involved everyone in the kingdom in some way. Courtiers and priests from the palace controlled the work, while engineers, craftworkers and thousands of peasants carried it out.

Giza in its heyday

The site at Giza originally contained much more than three huge pyramids. Each pyramid was linked to a chapel by a raised path called a causeway. After his death, a king was taken across the Nile by boat. He was then taken along the causeway and buried in his pyramid tomb. After the burial, a whole army of priests performed daily rituals to care for their king's soul. They thought the soul returned each day to his buried, preserved body.

Khufu

Khafre

Menkawre

The sphinx

Granite quarry, Aswan

This granite quarry is at Aswan, 500 miles south of Giza. The pyramid-builders used mostly soft limestone cut out of the ground at nearby quarries. However, the chambers inside Khufu's pyramid were lined with a polished stone called granite.

Transporting the stone

This recent photograph shows boats loaded down with large stone blocks on the river Nile. Stone used in the pyramids was also carried down the Nile by very similar boats. Some of the granite blocks for Khufu's pyramid weighed over 50 tons, the same as two large trucks.

The queens' pyramids

These three small pyramids were for Menkawre's three wives. Their tombs were never completely finished. Each king at Giza had smaller pyramids built next to his own for his queens.

Menkawre's queens' pyramids

Menkawre's pyramid

Khafre's pyramid

PREPARING A SITE

You will need: a stick (about 3 feet in length), modeling clay, string, tape measure.

THE shape of Khufu's pyramid at Giza is amazingly accurate. Making precise measurements in order to put up a building is called surveying. Modern surveying equipment produces results no better than the equipment used by the Egyptians 4,000 years ago. Building an accurate pyramid requires just two things—the outside of the base must be level and the blocks must be laid absolutely flat, with their sides facing vertically upward. The pyramid-builders used simple plumb lines to check that the verticals were straight. They also cut channels in the ground around the pyramid and filled them with water, to help them check that the base was level. These projects show you simple ways of using a plumb line and water levels to check verticals and levels.

A modern plumb line. A plumb line [i]s a weight on the en[d] of a line. The weigh[t] holds the line straigh[t]

Is it vertical?

1 Make your own plumb line by attaching a ball of clay to a piece of string. Make a large knot in the string and model the clay around it. The ball should be about 1 inch across.

2 Push your stick into the ground. Make it as vertical as possible (pointing straight upward) just by looking at it. Keep moving it slightly until you are sure it is straight.

3 Now use your plumb line to check how straight your stick is. Get a friend to hold the line next to the stick. Measure between the stick and line at the top and further down. If the measurements are the same, your stick is vertical.

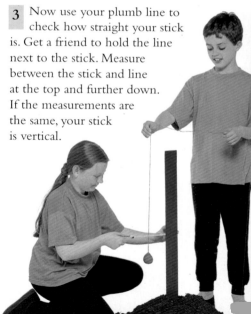

Is your site level?

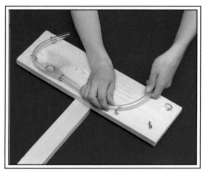

1 Here, you are making a simple version of a surveyor's level. To start, nail the short plank of wood to the top of the long wooden stick. Be careful with the hammer.

2 Now screw the hooks in along the bottom and up the sides of the plank. Thread the piping through the hooks. About 1 inch of piping should stick up above the plank.

3 Get a friend to hold the device upright. Now, very carefully, pour the water into the pipe, using a funnel. The water level should come just above the piece of wood.

4 Get a friend to hold the pole up, about 16 feet away. Look along the water levels so that they line up and also line up with the pole. Ask your friend to move a finger up and down and shout out when it lines up with the water levels. Get them to mark this on the pole.

5 Ask your helper to move to another spot. Repeat step 4. The distance between the two marks on the pole shows the difference in level between the two places.

MATERIALS

You will need: wooden stick 3 feet long, piece of wood 16 inches x 4 inches, nails, hammer, screw-in hooks, clear plastic tubing (about 1/8 inch across), pitcher, funnel, water colored with ink, tape measure, pen, wooden pole 3 feet long.

BUILDING THE GREAT PYRAMID

I MAGINE you are helping to build the Great Pyramid (Khufu's pyramid) at Giza. You take stone out of the ground by driving large wedges into cracks in the stone to split off smaller blocks. You haul the blocks more than half a mile from the quarry to the building site. You then have to lift them up more than 325 feet, to place them on the top layer of the growing pyramid. Each block weighs over two tons. In ancient Egypt, wooden sledges and rollers were used to move the massive stone blocks more easily. Large wooden poles were used as levers to put the blocks in position.

Nelson's Column

Sydney Opera House

Statue of Liberty

Great Pyramid, Giza

The Great Pyramid at Giza is the largest stone building in the world. It is much larger than some of the world's most famous huge landmarks.

Up to 12 strong men were needed to move just one block on its sledge. Building continued throughout the year, and at one point 20,000 laborers were working on this great pyramid.

FACT BOX

• Khufu's pyramid was built from more than two million blocks of stone. Each block is roughly the size of a kitchen table and weighs about 2.5 tons.

• The weight of Khufu's pyramid—around five million tons—is about the same as 250 large passenger cruise ships.

The mysterious Sphinx

A huge, mysterious stone statue called the Sphinx guards the pyramids at Giza. It has the body of a lion and a human head and is a form of the sun god. The Sphinx's head was carved from an outcrop of rock. More rock was dug from the ground around that area to form the body.

King Khufu's chamber

Khufu's body was placed in this chamber, inside his pyramid. It is lined with red granite and stands at the top of a sloping shaft—the Grand Gallery.

Inside the Great Pyramid. This is the only pyramid with passages inside as well as underneath. When workers finished the pyramid, they sealed off the Grand Gallery with stone blocks. They then left through the escape shaft.

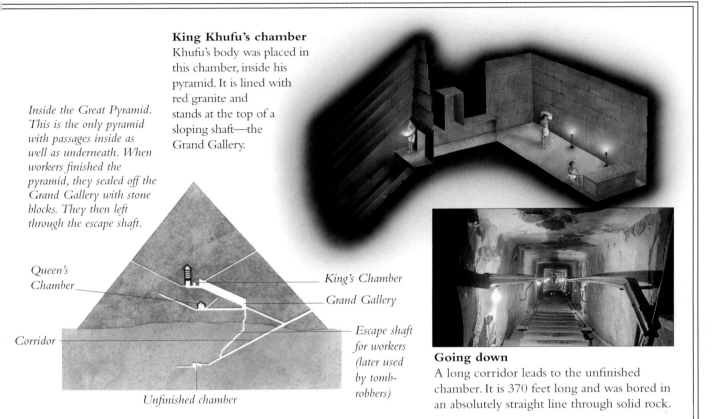

Queen's Chamber

King's Chamber

Grand Gallery

Corridor

Escape shaft for workers (later used by tomb-robbers)

Unfinished chamber

Going down

A long corridor leads to the unfinished chamber. It is 370 feet long and was bored in an absolutely straight line through solid rock.

Using ramps

This copy of a tomb painting shows that the pyramid-builders knew that a sloping ramp makes it easier to move large loads such as stone blocks. A straight ramp leading to the top of the pyramid was probably used. It would have been almost 1 1/2 miles long.

MOVING LOADS

You will need:
a short and long ruler,
3 square or rectangular erasers
(the same size), a load such
as a large potato.

Using levers

A LEVER is an extremely simple device that can be used to help people to lift all kinds of loads. The laborers working on the pyramids of ancient Egypt used levers made of lengths of wood to lift their enormous stone blocks into place. In the first project, you will find out how a lever can be used to turn a small amount of effort into a much larger force. In the second and third projects, you will learn about another way of making large loads easier to move. This is by cutting down on friction—a force that prevents things from sliding against each other. The ancient Egyptians used huge wooden rollers to cut down on friction and move their heavy loads along. Now you can try putting this to the test yourself.

Roller conveyor belts are used in all kinds of modern warehouses. Like the ancient Egyptians, we still use simple ways of reducing friction, such as rollers, to make loads easier to move.

1 Set up your materials. The eraser on the right is a fulcrum (pivot-point), and the long ruler is the lever. Touch the end of the lever lightly. You will lift the potato quite easily.

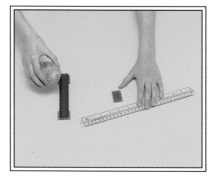

2 Now try it again with the fulcrum moved closer to the load. Even less effort is needed. The closer the fulcrum is, the less effort is needed to lift the load with the lever.

3 Now try lifting the load using just your fingertips. Can you feel how much harder it is without a lever? Experiment with different loads and levers and compare the results.

Reducing friction

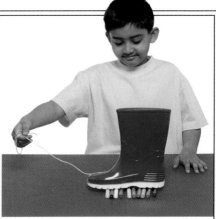

1 Pulling loads along is made easier or harder depending on the surface. Tape a length of string to your boot and pull it along a smooth surface (always ask permission first!).

2 Scatter sand on the surface. You can pull the boot over this more easily because the sand reduces friction. Each sand grain is like a tiny ball that rolls as the boot moves over it.

3 Place a row of round felt-tip pens on the surface and pull your boot over them. The boot will move even more easily over these rollers, as they are much larger than sand grains.

Feeling the force

1 Tie string around your package. Attach a rubber band. Make two marks on the band, about 1 inch apart. Pull the package along.

MATERIALS

You will need (Reducing friction project): something to pull, such as a rubber boot, tape, string, sand, round felt-tip pens.

You will need (Feeling the force project): some heavy books wrapped up to form a bundle, string, strong rubber band, felt-tip pen, ruler.

2 Discover how much the rubber band has stretched by measuring the distance between the two marks. This will give you an idea of how much force was needed to pull the load. Try it with lots of different loads, moving over different surfaces.

PYRAMIDS AND MUMMIES

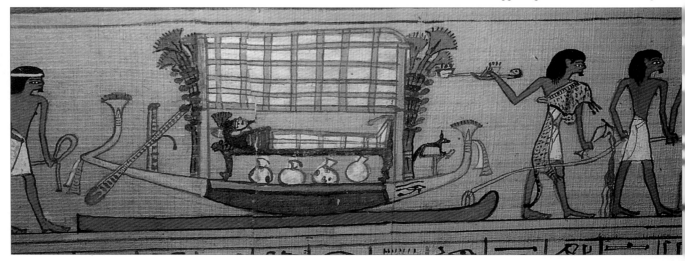

ACCORDING to the religions of ancient Egypt, the king was a god who had come down to live on earth. All through his reign he prepared for the end of his life, when he returned to heaven. After death, his body had to be preserved, as a "mummy," so that it would not decay. First, soft parts, such as the brain, were removed. The body was then buried in natron (a kind of salt). This dried out all the fluids that cause a body to decay. It was then wrapped in bandages and put in a wooden coffin. The king's ka (spirit) was said to move between heaven and his pyramid tomb. The Egyptians believed that, as long as his body remained preserved, the king would continue to visit the earth. His people looked across the Nile to the pyramids and felt that their king was still watching over them from his pyramid tomb and that they were safe and secure.

The area around a typical pyramid tomb looked like this. Close to each pyramid was a mortuary temple. It was linked by a covered path to a valley temple, close by the river's edge.

A boat that looked much like this was used to carry the king from his palace to the valley temple. Here, the body was dried out and wrapped up so that it did not decay.

Receiving offerings

This is a statue of King Khafre. It stood in the chapel linked to his pyramid. The Egyptians thought that, after death, a king's spirit re-entered his body and any statues of him. So priests made offerings of food and drink and believed that he could still enjoy them.

Canopic jar

The Egyptians used this type of jar to store soft body parts such as the stomach. These were removed from the body because they decay easily. This particular jar dates from some years after the age of the pyramids and is shaped like a jackal.

Protected by the gods

This painting dates from the New Kingdom era and shows a priest in a jackal mask, attending to a mummy. The mask that the priest is wearing represents Anubis, the god of death, who protected the tombs and the mummies inside them.

Turning the body of a great Egyptian king into a mummy was an extremely elaborate procedure. At each stage of the process, priests and embalmers said special prayers and followed particular sacred rituals.

PRESERVATION

This photo was taken in 1909. It shows the mummy of King Seti, who died more than 3,000 years ago. Archaeologists have cut off most of the linen bandages. You can see clearly just how good the ancient Egyptians were at preserving bodies.

MATERIALS

You will need: rubber gloves, 2 apples, 2 carrots, peeler, plastic garden tray with holes, newspaper, soil mix, stone, plastic object, piece of wood, spade.

PRESERVING something means stopping it from decaying. The Egyptians knew all about this and preserved their kings' bodies for many years. Decay is caused by invisible creatures called bacteria and tiny funguses called mold. Bacteria are everywhere. While we are alive, our bodies fight them. As soon as something dies, bacteria and mold start to cause decay. Bacteria grow best in moist places, so the Egyptians made a mummy by removing moist body parts and drying out the rest. Mummies were also wrapped in bandages soaked in oily resins that killed bacteria and molds—like modern antiseptic creams. Find out more with these projects.

What kind of things decay?

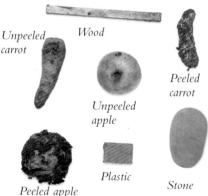

Unpeeled carrot *Wood* *Peeled carrot*

Unpeeled apple

Peeled apple *Plastic* *Stone*

1 Peel one of the apples and one of the carrots. Line the tray with some newspaper. Add a layer of soil mix and place all the items on top. Add more soil mix to cover the items.

2 Dig a shallow hole in a shady spot and put the tray in it. Cover it with earth so that you can just see its top edges. Buried like this, the items will stay damp. Dig it up after a week.

3 Examine the results. Fruit and vegetables are attacked quickly by bacteria and mold, especially if they have no skin. Wood takes months to decay. Stone and plastic do not decay.

Preventing decay

1 Put one slice of ordinary bread into a plastic bag and seal it with a tie. Now toast another slice of bread until it is crisp and dry. Seal the toast in another plastic bag.

2 Spread some antiseptic cream, which is designed to kill germs, over one side of a third slice. The Egyptians used bandages soaked in oily resin to wrap their mummies.

3 Label each plastic bag. Set the bags aside in a warm place and check them once a day. Bacteria and mold are everywhere. What will they do inside your plastic bags?

4 Look at the slices. Mold and bacteria cannot grow on toast, as there is no moisture. The cream on the second slice has killed any germs, and the ordinary slice is very moldy.

Dry toast is not affected.

Chemicals in the cream kill germs.

Ordinary bread gets very moldy.

MATERIALS

You will need: 3 slices of bread, 3 plastic bags with ties, 3 tags, gloves, pen, knife, antiseptic ointment.

5 Do not open the bags when you have finished looking at the results. Keep the mold and bacteria wrapped safely inside their plastic bags and drop the bags into a trash can.

ANCIENT EVIDENCE

THE walls inside the pyramids at the Giza site are completely bare. There are no pictures, carvings or writing. The first ancient Egyptian king to have writing on the walls of the burial chamber inside his pyramid was Unas. King Unas lived about 250 years after Khufu (Khufu became king in 2589 B.C.). The writings that appear on the walls are known as the Pyramid Texts. These texts are not written in letters from an alphabet. Instead, the Egyptians used a kind of writing called hieroglyphics. In hieroglyphics, each word is represented by a picture. In later years, all Egyptian tombs were covered with these texts. Toward the end of the ancient Egyptian civilization, even the tombs of ordinary people had a small pyramid on top and texts on the walls. These texts tell us all kinds of interesting things about just what everyday life was like in the land of the great pyramids.

In 1954, Khufu's funeral boat (above) was found in a pit (below) alongside his pyramid. The Egyptians thought his soul traveled across the sky in a boat.

Unas's pyramid
The walls of this underground room in Unas's pyramid are completely covered with hieroglyphic writing. It tells the dead king how to travel into the sky and meet the sun-god Re.

The Rosetta stone

This flat stone slab is called the Rosetta stone. The same message appears on the stone, but it is written in Greek and in two types of hieroglyphics. In 1822, Jean-François Champollion cracked the Rossetta stone code. The Greek on the stone enabled him to translate the hieroglyphics.

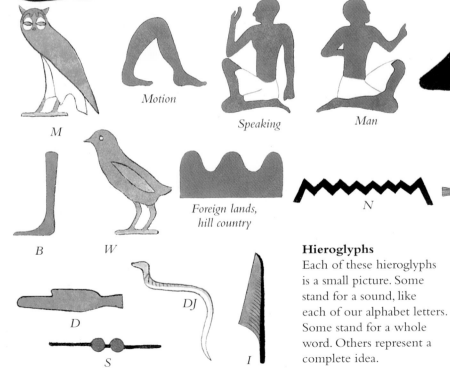

M

Motion

Speaking

Man

Q

K

T

H

B

W

Foreign lands, hill country

N

Life

D

DJ

S

I

Scribe (learned person)

Hieroglyphs

Each of these hieroglyphs is a small picture. Some stand for a sound, like each of our alphabet letters. Some stand for a whole word. Others represent a complete idea.

USING SEALS AND PAINTS

MATERIALS

*You will need:
modeling wax, warm water,
potato, old knife,
felt-tip pen, envelope.*

Yᴏᴜ have seen some of the beautiful pictures and hieroglyphs that the ancient Egyptians painted on their walls, including some of their pyramids. In the second project shown here, you can make paints in just the same way and produce some of your own pictures. Perhaps you would like to paint your own scene of life in ancient Egypt, or copy one of the hieroglyphs shown on the previous page. Another thing that the ancient Egyptian people used was seals. They placed seals on the doors of tombs and on chests, so that people could not get in easily. Their seals were made of mud. Many years later, people around the world were using seals on private letters, but these were made of wax. In the first project here, you can learn how to make patterns in seals and how seals work. This means that you can put an attractive seal on your own letters.

Experiment with different designs for your personal seal.

Make your own seals

1 Drop the wax into warm water for a few minutes. Squeeze it with your fingers until it is very soft. Let it sit in warm water while you do the next step.

2 Cut the potato in half. Draw a simple design on one half. Now use the knife to carve out your design. The design should be in relief (so it stands out), as shown in the close-up.

3 Take the wax out of the water. Drop it quickly onto the envelope. Press the potato into it and lift it off to reveal your seal.

Ancient painting methods

1 Paint color comes from powders called pigments. The Egyptians used powdered minerals. Use colored chalks and charcoal to make your pigments. Crush them on a plate with a spoon.

2 Paint is made by mixing pigment with a binder like egg white (the Egyptians used this or gum). Stir egg white, water and pigment together until you have a thick, strong paint.

3 Now make your brush. Cut pieces of straw and make them into a bundle about ¼ inch thick. This is the kind of material the Egyptians used for their paintbrushes.

M A T E R I A L S

You will need: colored chalks, charcoal sticks, spoon, egg, 2 small dishes, water, straw, rubber band, pen, scissors, paper.

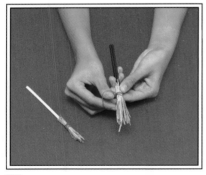

4 Use a rubber band to attach the bundle of straw to the end of a pen or pencil. Trim the brush with scissors to make the tip pointed. Make more brushes of different sizes.

5 Now use your paints and brushes to create a picture. To be as much like an ancient Egyptian picture as possible, your painting should be in browns, yellows, black and white.

Try painting pictures of Egyptian hieroglyphs.

31

PYRAMIDS AND THE STARS

This painting is from the ceiling of a king's tomb. It shows several important Egyptian gods and their stars in the night sky.

THE three enormous pyramids at Giza hold many secrets. Even though they are empty, they are filled with hidden meanings. Most of these meanings are linked with stars in the sky. The Egyptians thought that each star was actually a different god. When a king died, his spirit returned to the sky and lived there in the form of a star. The three Giza pyramids are laid out on the ground in the same pattern as the stars known as Orion's belt. Also, four passages in Khufu's pyramid pointed exactly at four stars that represented the four most important Egyptian gods. It seems that the ancient Egyptians may have known a great deal about the stars.

Finding north and south by the stars

Before the building of a pyramid began, a priest set the north-south line by using a curved wall and a forked stick. He noted where a star rose above the wall and where it set, farther along the wall. The north-south line ran from the stick to a point midway between the star's rising and setting. The base of Khufu's pyramid is laid out facing exactly north-south and east-west. It is so exact that you would think it had been laid out with modern instruments.

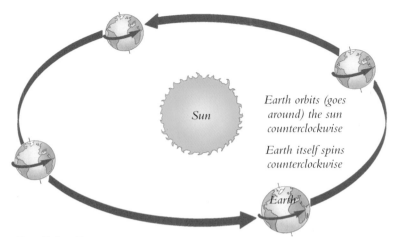

View of our spiral galaxy   *Top view*

x = *Sun*   *Side view*

Our Galaxy

Our sun is actually a star. There are billions of other stars, arranged like a spinning Catherine wheel called the galaxy. Our sun is near the rim of the galaxy. There are billions of other galaxies in the universe, far away from ours. We cannot see the stars during the day because the sun is so bright.

Earth orbits (goes around) the sun counterclockwise

Earth itself spins counterclockwise

Our Solar System

The earth spins around our sun, giving us a changing view of the space around us. The earth itself also spins around once a day, so the sun seems to move across the sky. At night, the stars appear to move in curved paths. Some of the brightest points of light are the planets Mars, Venus and Jupiter, lit by the sun. It may be that the ancient Egyptians used these to find the north-south line.

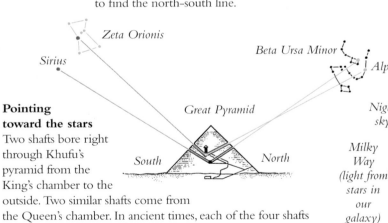

Pointing toward the stars

Two shafts bore right through Khufu's pyramid from the King's chamber to the outside. Two similar shafts come from the Queen's chamber. In ancient times, each of the four shafts pointed directly to certain stars—home of important gods and goddesses acccording to Egyptian thought. The pattern of stars in the sky has changed since the pyramid was built, but computers can show how the stars appeared in Khufu's time.

The pyramids and the stars

Some people have suggested that, at Giza, the Nile forms a similar shape to the Milky Way in the night sky overhead. Also, the three Giza pyramids are arranged a lot like the three stars in the middle of a group of stars called Orion's Belt. The Egyptians may have been trying to make a place on earth for their gods, who lived in the stars.

FINDING NORTH AND SOUTH

You will need:
globe, toothpick, modeling clay or
reusable adhesive, lamp, masking
tape, scissors, other tape (optional),
tape measure.

As we have seen, Khufu's pyramid at Giza was constructed in an amazingly accurate way. Its massive base is almost a perfect square, and the sides of this base point exactly north-south and east-west. The ancient Egyptians did not have magnetic compasses to find north and south as we do today. Instead, they used the stars. In these two projects, you can use the Sun in exactly the same way. In any case, our sun is just a very large star. First of all, you will use a model to help you understand some basic principles. Then you can try it for real outdoors. In the first project, the first shadow that you mark represents the shadows that you see around you in the middle of the morning. The second shadow represents shadows that fall mid-afternoon. The lamp represents the sun, which casts these shadows. As you turn the globe around, this represents how the earth rotates every day.

Finding north

1 Stick the toothpick vertically on Egypt. Switch on the lamp (the sun). Turn the globe until a shadow appears. Stick a piece of tape under the shadow.

2 Turn the globe until you get another shadow of the same length, pointing in the opposite direction from the first one. Mark this with tape.

3 Now cut some tape and stick it down so that it joins the ends of the other pieces. Use a tape measure to find the center of the third tape strip.

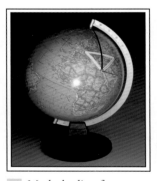

4 Mark the line from Egypt (the base of the toothpick) through the middle of the third strip with more tape. This points exactly to the North Pole.

Moving outside

1 Choose a bright sunny day for this project. Around mid-morning, push a stick vertically into the ground. Use the plumb line you made earlier to check that it is vertical.

2 The stick will cast a shadow on the ground. Cut two paper strips exactly the same length as the shadow. Place one on the ground directly under the shadow.

3 As time goes by, the shadow will move and change length. Later in the afternoon, it will be the same length as the second paper strip. Mark this shadow with the strip.

M A T E R I A L S

You will need:
stick (3 feet long), plumb line,
2 long paper strips, tape measure,
string, 2 large stones, scissors.

4 Just as you did with the model globe in the first project, join the two ends of the paper strips. One effective way of doing this is to use string wrapped around and held down with stones.

5 Use the tape measure to find the middle of the length of string. Join this point to the bottom of the stick. This line points exactly north-south, just like it did on the model globe.

NORTH AND SOUTH AMERICA

Apart from Egypt, the other large area of the world where pyramids are found is Central America. This is the area from Mexico down to the northern part of South America. Ancient Egypt was cut off from its neighbors by deserts and the sea, and just one distinct group of people lived there. In Central America, however, there were many different tribes of people, such as the Olmecs, Toltecs, Maya and Aztecs. Over thousands of years, these tribes moved back and forth, fighting and conquering each other. The pyramids in this region are less than half as old as the ones in Egypt—most were built between 1,500 and 500 years ago. Unlike Egyptian pyramids, pyramids in the Americas have wide staircases running up the outside to a temple at the top. In Egypt, the pyramids have stood undisturbed in the dry desert. In Central America, explorers have often found pyramids half-buried, deep inside hot, steaming jungles.

Most of the pyramids are in Central America. This is the area where the North American continent meets the South American continent.

Sun pyramid, Teotihuacán
This pyramid of the sun was built around A.D. 300 at Teotihuacán, near modern Mexico City. It has a solid core made from millions of sun-dried mud bricks. The outside is covered in blocks of hard stone.

			LAST EGYPTIAN PYRAMIDS	500		B.C. ◄—	0 —►	A.D
MEXICO								
YUCATAN PENINSULA								
NORTH AMERICA						HOPEWELL	PEOPLE	

Pyramid of the Magician, Uxmal, Mexico

This pyramid was built somewhere between A.D. 500 and A.D. 900, on Mexico's Yucatán Peninsula. An impressive main stairway leads up to a large temple perched on the top. This pyramid is particularly unusual because it has curved walls. Uxmal was once a prosperous town, filled with grand buildings, which flourished between about A.D. 600 and A.D. 1000. The city was totally abandoned in the 1400s.

Temple of the Inscriptions, Palenque

This was built by the Mayans in southern Mexico about 1,200 years ago. Like many pyramids, it has a stairway and a temple at the top. It is one of very few pyramids in this part of the world that was used as a tomb. The actual sarcophagus (coffin) had a beautiful lid, which you can see later on in the book.

Tula, Mexico

This pyramid is at Tula, the capital of the Toltec empire. Like most Central American pyramids, it does not rise to a point. A temple once stood on its top.

500	1000	1500
CITY OF TEOTIHUACAN	TOLTECS	AZTECS
MAYA		

WORKING WITH STONE

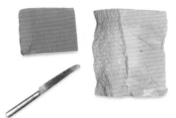

*You will need:
blunt, round-ended knife,
special block used by
flower-arrangers, foamed
concrete (fairly soft) type
of building block.*

THE major pyramids were made from stone (rock) because this lasts much longer than earth or bricks. This meant that a huge amount of stone had to be cut out of the ground and carved. Different types of stone produced a variety of effects. There are three main types of rock—sedimentary, igneous and metamorphic. Sedimentary rocks formed long ago, when particles of sand or tiny animal shells built up in layers under the sea. Igneous rocks formed near volcanoes when molten rock cooled and became solid. When these rocks are later heated underground, they change into metamorphic rock. Some rocks are hard, while others are much softer. Certain rocks have beautiful patterns and colors in them. The Central Americans used a lot of granite, an igneous rock, in their buildings. The ancient Egyptians used lots of limestone, which is sedimentary. Both carved stone with simple hammers and chisels. These projects help you to take a closer look at rocks and discover some of their properties.

Now you know how hard it is to carve material that is much softer than stone. How long would it take you to carve stone with simple tools?

Cutting and carving

1 Could you be a sculptor? Practice by marking out a simple shape on the flower-arranging block. Use the knife to cut around it. How easy is it to use this material?

2 Now do the same with your building block. This is much harder, but it is still softer than stone. Make sure that the knife blade is pointed away from you.

Learning about stone

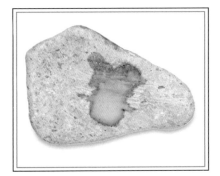

1 Do this vinegar test to give yourself some clues about what your stone samples might be. Put a few drops of vinegar on each rock sample and then watch carefully.

2 Limestone *(seen here)*, chalk and marble make vinegar fizz. They react because they are types of calcium carbonate. The common rocks—flint, granite and sandstone—are unaffected.

3 Now scratch one sample with another. Harder rocks leave marks on softer ones. Hardness also helps you to identify rocks. Think how difficult it might be to carve the harder ones.

Flint

Granite

Limestone

Sandstone

MATERIALS

You will need: vinegar, spoon, samples of different stones, magnifying glass, reference book.

You can see small crystals in granite.

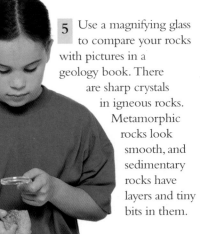

5 Use a magnifying glass to compare your rocks with pictures in a geology book. There are sharp crystals in igneous rocks. Metamorphic rocks look smooth, and sedimentary rocks have layers and tiny bits in them.

4 Arrange your rocks in order of hardness. Igneous rocks, like granite, are usually the hardest. Sedimentary rocks, like sandstone, are usually the softest.

THE MAYAN PEOPLE

THE Maya were a nation of people who lived in the Yucatán peninsula area of Central America between about A.D. 250 and A.D. 1000. They built spectacular pyramids, but they rarely buried their kings inside them. Mayan pyramids were aligned north-south very accurately, and from the tops of their pyramids, the Maya made careful measurements of how the sun, moon and stars moved. The Maya were fascinated by time and drew amazingly complicated calendars, based on their observations of the stars. Our calendar covers 365 days and has an extra day every leap year. One of their calendars covered 52 years and followed the stars more accurately than ours. Every day was linked to a different god. Each god required its own special prayers and offerings. Various things brought about the end of the great Mayan civilization. One of the main reasons was probably fighting between the rulers of the different city-states.

This carved head shows how Mayan people looked long ago. Many of the people living in Central America today have very similar features.

FACT BOX

• The Mayan use many different calendars; a 260-day calendar is still used in some remote regions of Central America today.

• Inside the famous El Castillo pyramid is another, smaller pyramid. This inner pyramid was constructed 100 years before the one we see today. It was probably built by another group of people called the Toltecs.

Grand Plaza, Copán, Honduras
This pyramid is at Copán, one of the main places used by the Mayan astronomers. Like the Egyptians, they believed that life on earth is controlled by the way the stars move. They believed that their calendars could predict the future.

Palenque, southern Mexico

The Temple of the Inscriptions stands on the top of this pyramid. Its walls are covered with strange writing and carvings. In 1949, workmen removed a stone slab from the floor and found a stairway. It led down inside the pyramid to a burial chamber. Here they found the remains of six people who had been sacrificed and a great carved stone coffin with a beautiful lid *(see below right)*.

Tomb lid from the Temple of the Inscriptions

Here you can see the lid of the coffin buried under the Temple of the Inscriptions. The lid weighs 5 tons. Its carvings show a Mayan nobleman surrounded by sacred symbols such as a dragon. These carvings have taught us many things about how Mayans lived.

El Castillo, Chichén Itzá, Yucatán Peninsula

This magnificent pyramid is the largest pyramid in the Mayan city of Chichén Itzá. It was built around A.D. 1100. The number of steps up the sides and in the temple equals 365—the number of days in a year.

THE CITY OF TEOTIHUACAN

Teotihuacan is an enormous ancient city that lies 25 miles to the northeast of Mexico City. At the end of the 1800s, it was a crumbling mass of overgrown ruins and half-buried pyramids. Today, however, it is one of the world's greatest ancient sites—hundreds of thousands of tourists visit it every year. Around A.D. 500, Teotihuacán was one of the largest cities in the world, with about 120,000 people living there. About one third of the population were skilled craftspeople, and there were more than 800 workshops supplying tools, knives, statues and pots to the city and surrounding countryside. All the city streets were laid out on a square grid, just like modern New York City. Running right through the center of Teotihuacán was a wide road called the Street of the Dead, which we can still see today.

The Street of the Dead
This road runs through the center of the city. At one end is the pyramid Temple of Quetzalcoatl. At the other is the Pyramid of the Moon. Halfway along is the huge Pyramid of the Sun.

Staircase, Temple of Quetzalcoatl
This stairway, running up the side of the Temple of Quetzalcoatl, is lined with carvings of jaguar heads. Jaguars represented the fertility of the earth. Rulers wore jaguar skins as symbols of their power.

FACT BOX
• An engineer studied Teotihuacán very closely during the 1960s. He decided that the Street of the Dead was actually a scale model of the solar system. This is what he thought it all meant:

– The Temple of Quetzalcoatl is the sun.

– Markers along the street represent the inner planets Mercury, Venus, Earth, Mars and Jupiter.

– The Pyramid of the Sun represents the planet Saturn.

– The Pyramid of the Moon is Uranus. (Note that Uranus was unknown to Western astronomers until 1787.)

Sizes of planets are shown to scale. Distances of the planets from the Sun are not shown to scale.

Sun	Mercury	Venus	Earth	Mars	Jupiter	Saturn	Uranus	Neptune
	35	65	90	140	485	870	1800	2800

DISTANCES ARE SHOWN IN MILLION MILES

Our solar system consists of the sun and the planets. Venus was particularly important to the pyramid astronomers of Central America. Some archaeologists have thought that the Street of the Dead, in the ancient city of Teotihuacán in Mexico, mirrors the solar system.

Pyramid of the Moon

From the top of this pyramid, there is a magnificent view over the city and of the Street of the Dead. Like all the buildings at Teotihuacán, the pyramid was once painted with patterns and scenes. Imagine how bright and colorful the city must have been in its heyday.

Pyramid of the Sun

The base of the Pyramid of the Sun is 700 feet long. It is 230 feet high and its volume is 35 million cubic feet. The mud used to make the sun-dried bricks would have filled 5,000 Olympic-sized swimming pools. The pyramid was built in four huge steps. At the corner of each step, excavators found the skeleton of a person who had been buried alive.

THE TOLTECS

THE Toltecs were a warring race. The great days of their civilization followed after most of the Mayan empire had started to decline. They attacked and looted Teotihuacán and the surrounding lands in about A.D. 750. As time went by, the Toltecs repaired old pyramids and temples and built many new ones. They used stones or bricks made from sun-dried mud for the insides and covered the outsides with slabs of smooth stone. Just like all Central American pyramids, there were staircases up the sides and temples at the top. It seems that Chichén Itzá, in Mexico, probably became a major Toltec city, as it had been for the Mayans. It is famous for the huge pyramid of El Castillo. As the Toltecs conquered more places, they became more ferocious. They believed that their gods needed blood, and thought that human sacrifice was the only way to keep the sun burning. The sacrifices were made in the temples at the tops of the pyramids.

The major city of Chichén Itzá, 1,000 years ago. Its main building was the pyramid El Castillo, which towered over the main square. The Toltecs had great influence over the area around Chichén Itzá, and may have ruled here.

Chacmool figure

At the top of El Castillo was a stone figure called a Chacmool. These figures were found on various sites in the Americas. Each Chacmool has a plate on its belly. The Toltecs believed that their god Quetzalcoatl had sacrificed his heart and blood to make the sun. To keep the sun burning, Toltec priests cut out the hearts of human victims and piled them onto the Chacmool's plate.

44

Temple of the Warriors, Chichén Itzá

This temple is surrounded by 60 carved columns. The carvings show the weapons used by Toltec warriors. The warriors were split into groups—the Eagles, Jaguars and Coyotes—and wore huge, feathered headdresses.

Caracol observatory, Chichén Itzá

The Caracol observatory (a building for looking at the night skies) was built by the Mayans but would also have been used by Toltecs. Its windows and pillars line up with the planet Venus. To the Mayans and the Toltecs, Venus was closely linked to the sun, which was also extremely important to their beliefs.

The Cenote of Sacrifice, Chichén Itzá

This massive well measures 200 feet across. Steep, rocky sides drop down to murky water 65 feet below. All kinds of offerings—even people—were thrown in here. In 1962, divers brought up over 4,000 objects, including gold discs that showed various scenes from Mayan and Toltec religious rituals.

HOW BIG?

You will need: 4 large stones, 4 balloons to tie to stones to make clear markers, compass, tape measure, string, scissors.

MOST of the world's pyramids are huge. Remember that the largest—the Great Pyramid at Giza—is 480 feet high and the base length is 750 feet. It can be difficult to imagine how great these numbers are. Each side of the Great Pyramid is about 2¼ times longer than a 100-yard race track and 1½ times taller. It covers an area of 13 acres—the same as nearly seven soccer fields. These two projects concentrate on the El Castillo pyramid at Chichén Itzá. They help you understand the gigantic size of pyramids. In the first project, you set out the actual size of the pyramid's base. In the second, you make a scale model.

Use your feet to measure out distances in the first project. To do this, find out how many foot-lengths make up 1 yard.

A life-size pyramid

1 Find an open space. Put down a stone marker. Using a compass, walk due north. Take the number of heel-to-toe steps that make up 55 yards (or 180 feet) —the length of one side of El Castillo.

2 When you get to this point due north, put down a second stone marker. Now set off again, taking exactly the same number of steps—due east this time. Place another marker and set off due south. Place another marker and set off west, coming right back to where you started. You have now plotted the base of El Castillo.

El Castillo, Chichén Itzá

The project opposite gave you some idea of just how big El Castillo is. This recent picture shows how big it is compared to people. However, it is still much, much smaller than the Great Pyramid at Giza! In this second project, make a scale model of El Castillo and compare it with a model car.

Scale model of El Castillo

1 This model car is 2 inches long. A real car might be 11½ feet long. We say that the scale of the model car is 1:70. In other words, the real car is 70 times bigger.

M A T E R I A L S

You will need: model car, tape measure, 4 stones, string or line of some sort, scissors.

2 Make a 1:70 scale model of El Castillo, as shown. The base should be 31 inches long (2200 ÷ 70). The sloping sides should measure 16 inches long. Use large stones to keep the pyramid base in shape.

3 Pull up the strings that form the sides. Your car was made to the same scale. Place it next to the pyramid to see how large a car would look next to El Castillo. Could you make a model of Khufu's pyramid? (If you made a model of Khufu's pyramid to the same scale, its base sides would be 10 feet long and its height would be 7 feet.)

THE AZTECS

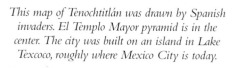

THE Aztec people came after the Toltecs. Their nation was at its peak between about A.D. 1250 and A.D. 1519. The Aztecs were the last major civilization in Central America and the last people to build pyramids in this part of the world. Their capital city was Tenochtitlán. In 1519, soldiers from Spain invaded and swiftly defeated the Aztecs, and so Spanish is now the main language of Central America. Like the Maya, the Aztecs had calendars based on the movement of the stars. Priests studied the calendars and predicted eclipses of the sun and moon and the end of the world. Like the Toltecs, they believed that the only way to stop this end was to sacrifice humans. Many of these sacrifices took place in their pyramid-temples. The Spaniards disapproved of this and tried to convert the Aztecs to Christianity.

This map of Tenochtitlán was drawn by Spanish invaders. El Templo Mayor pyramid is in the center. The city was built on an island in Lake Texcoco, roughly where Mexico City is today.

FACT BOX

• El Templo Mayor pyramid in Tenochtitlán was built in six layers. One new layer was added every 52 years—a significant time in the Aztec calendar.

• Aztec myths said that the god Quetzalcoatl would return from the west as a pale-skinned, bearded man. Cortés, the Spanish conqueror, fitted this description, so his conquest of the Aztecs was made far easier.

Cortés and Montezuma, in 1519
This Spanish painting shows the Aztec leader Montezuma welcoming the Spanish leader Hernándo Cortés to Tenochtitlán in 1519. Montezuma did not fight the invaders, but simply gave his throne to Cortés.

Stone skull rack

This stone carving shows a rack of human skulls. Each pyramid temple had a real skull rack for holding the heads of sacrificed prisoners. The Spanish invaders were horrified to find a rack at Tenochtitlán that held around 10,000 skulls. They thought the Aztec people were totally barbaric.

Stone statues at El Templo Mayor

Each of these stone statues has a hole in its chest in the heart area. This shows that the Aztecs removed victims' hearts when they sacrificed them.

Modern Mexico City

Mexico City, the capital of Mexico. Tenochtitlán's ruins lie buried under the modern city. Can you see the cathedral tower near the center? The cathedral was built partly with Aztec stones. It was completed in the 1800s.

MODERN PYRAMIDS

Have you ever seen a modern pyramid-shaped building? Most of them look like pyramids on the outside but, inside, they are just like ordinary buildings. Ancient pyramids were built from earth or stones piled up to make huge mounds. Modern pyramids are much smaller and lighter. Instead of huge stone blocks, they are made from concrete, steel and glass. Inside modern pyramids, there is a steel frame. This works like a strong skeleton, supporting concrete slabs that make up the outside walls. There is something about a pyramid shape that still catches people's attention and impresses them. People today, just like the people of ancient Egypt or Central America, often choose a pyramid when they want something special.

Forge shopping center, Glasgow
This pyramid-shaped shopping center is in Glasgow, Scotland. Today there are quite a few buildings all over the world, such as offices, banks and hotels, that are pyramid-shaped.

The Louvre, Paris, France
This glass pyramid was completed in 1989. It was built around the entrance to the famous Louvre museum. Its sides slope at the same angle as the Great Pyramid at Giza. In this picture, you can see the pyramid reflected in water at night time. Many people admire this bold new building, even though it is so different from the older museum next to it.

Luxor Hotel, Las Vegas

Inside this pyramid you will find a hotel and gambling casino, complete with waiters dressed as Egyptian pharoahs. Covered in black glass, its main framework is made from steel girders. In front of the hotel is a copy of the famous Sphinx, but this one is made of plastic and concrete. The hotel was built in 1993 by an American businessman.

Transamerica building, San Francisco

On the left of this photo is the Transamerica pyramid skyscraper, which is 840 feet high. San Francisco is an area where there is a danger of earthquakes. A pyramid is a stable shape. The Transamerica building's pyramid-shaped steel frame helps it to withstand earth tremors.

Sadat Memorial, Cairo, Egypt

The Egyptian president Muhammad Anwar al Sadat was murdered in 1981. This pyramid was built as a memorial to him.

FACT BOX

• In a building shaped like a glass pyramid, warmth from the sun is trapped inside, just like a greenhouse. This helps to keep the building warm naturally.

• Glass pyramids must be fitted with outside rails on which mountaineering window-cleaners can clip their safety harnesses.

USING PYRAMID SHAPES

IMAGINE that you are standing up in a bus or train that is shaking as it moves along. How do you stop yourself from falling over? You probably spread your feet apart so that you are wider at the bottom than at the top—just like a pyramid shape. This shape is very stable and rigid, which means that it does not easily bend or buckle. High tension towers are extremely good examples of a modern use of pyramid shapes. They have a square base and sloping sides, which support the heavy cables that carry electricity to our homes, offices and factories. The pyramid shape of high tension towers means that they are able to stand up to hurricane-force winds. Pyramid shapes also help to hold up buildings. The Transamerica pyramid on the previous page is just like a huge high tension tower covered in concrete. From towers to skyscrapers to egg cartons—the pyramid is a useful and important shape that is used for all kinds of things in modern life.

This was the symbol for the British television station, Anglia TV. The pyramid shape is very effective because people recognize and remember it easily.

High tension tower
The tapering pyramid shape of a high tension tower makes it very stable. It is designed to use the least amount of metal to create the greatest possible strength. The electricity cables are attached to long glass insulators that hang down from the steel arms of the tower.

FACT BOX
• A large egg carton made of papier mâché *(see egg carton shapes on the opposite page)* will probably support your weight if you step on it—very carefully. This is because the pyramid shape is so strong, rigid and stable.

• Large high tension towers are about 100 feet tall.

• One of the first trademarks ever used was a red, triangular pyramid shape. It was used by a brewery.

Egg carton shapes

Egg cartons consist of paper or plastic pyramid shapes arranged side by side. They can hold any size of chicken's egg. Each egg slips downward until it is held by the sloping sides of the four pyramids surrounding each hole. Trays of eggs can be stacked one on top of another without breaking a single egg.

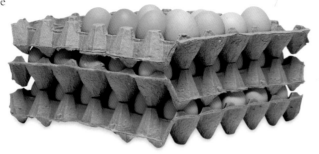

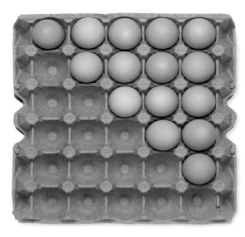

Obelisk shapes

This set was built in 1950 for the movie *Cleopatra*, about a famous queen of ancient Egypt. Here, you can spot another kind of traditional ancient Egyptian pyramid shape. This is the obelisk—a tall column with a small pyramid at the top. These were originally used as monuments or for religious purposes.

World War II tank traps

These concrete, flat-topped pyramids were placed in open ground during World War II. The idea was that the tracks of enemy tanks would slip down the sides, trapping the tanks.

BUILDING WITH PYRAMIDS

You will need:
cardboard or thick paper,
scissors, pencil, white glue, ruler,
weight (such as a
small package).

THE pyramid is a versatile, strong shape, so there are many uses for it today. Some of these have already been shown. Just as in ancient times, pyramids are still used in buildings. For example, in many parts of the world there is a danger of earthquakes. These make the ground shake violently and can cause buildings to topple over. Architects (the people who design buildings), have found that pyramid-shaped frames inside buildings can help prevent them from collapsing. The first project shows you some more proof of how strong and stable pyramid shapes can be. It also shows you how you can fit pyramid shapes together to form different shapes—and then discover what properties these new shapes have. In the second project, you will compare pyramids with box-shapes of exactly the same height. See for yourself which one is better at resisting violent shaking.

Learning about shapes

1 Make 12 identical cardboard pyramids. The sloping faces must be angled at 45°. To do this, make sure that the pyramid's height equals half the diagonal measurement across its base.

2 Make one big pyramid from 6 small ones, as shown. (There will be gaps around the edges.) Then make one big cube from 6 small pyramids, gluing the pyramids together.

3 Now try balancing a weight on your big pyramid and cube. They are equally strong. The pyramid shape is strong even though it has gaps in it. The cube is strong (unlike the cube on page 9) because it is made up of pyramid shapes.

Make a mini-earthquake

1 Make a small cube and a pyramid from modeling clay (choose a type of clay that is not too sticky). They must both be the same height and their bases must be the same size.

2 Place the cube and the pyramid on a book. Slowly tilt the book, to imitate the effect of an earthquake. The cube topples over long before the pyramid does.

3 The shapes you have just tested were low to the ground. Now try making two taller shapes. As before, they must both have the same height and base size.

4 Do the book test, as before. You should still find that the pyramid is the most stable. It is the second of the two shapes to fall over.

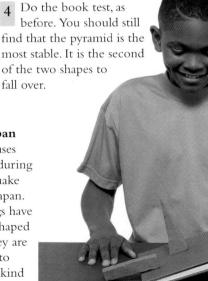

Kobe, Japan
These houses collapsed during an earthquake in Kobe, Japan. If buildings have pyramid-shaped frames, they are less likely to suffer this kind of damage.

PYRAMIDS AND CHEMISTRY

Some of the pyramids that you have seen so far—such as the Great Pyramid at Giza, in Egypt—are among the largest manufactured objects in the world. However, there are other pyramids that are much, much smaller and are found where you might not expect to find any pyramids at all. Some of these pyramids are so tiny that you need a magnifying glass or a microscope to see them. These objects are called crystals. You can see them for yourself in some everyday substances. For example, look closely at a spoonful of sugar or salt and you will see thousands of tiny crystals, each one shaped like a cube. Other crystals are shaped like tetrahedrons or square pyramids. They are often found deep under the ground, inside minerals (rocks). These minerals were formed when molten rock cooled slowly and changed from a liquid into a solid. As the rocks cooled, the crystals grew larger. These crystals are not just beautiful to look at. They also have many practical uses, as you will see.

This crystal of calcite is shaped like two square pyramids joined together at the base. For this reason its shape is called a bipyramid.

Quartz
This picture shows crystals of the mineral quartz. Tiny slivers of these pyramid-shaped crystals are used in quartz clocks and watches. The quartz controls the speed at which an electric current switches on and off. This in turn controls how fast the hands turn around.

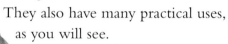

FACT BOX

- All the crystals of a particular substance normally have the same shape.

- There are seven main types of crystal shapes. These are called habits.

- Diamonds are a type of carbon. Pencils are made from another type of carbon.

- There is a form of carbon within all living things.

- Diamonds formed when carbon was heated and squeezed deep underneath volcanoes.

Sulphur

Sulphur crystals grow inside the craters of volcanoes. There are two types of sulphur crystal: dipyramid-shaped orthorhombic sulphur and column-shaped monoclinic sulphur. Millions of tons of sulphur are used each year to make sulphuric acid. This in turn helps to make plant fertilizers, paint, detergents and some plastics.

Volcanoes and crystals

The smoke rising from this volcano contains sulphur gas. This comes from deep underground. As the gas touches a sulphur crystal, it cools, solidifies and builds up an extra layer on the surface of the crystal.

Diamonds

These diamonds have been cut and polished to form beautiful gemstones. The carbon inside diamonds is arranged in tetrahedral patterns.

GROW PYRAMID CRYSTALS

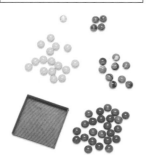

*You will need:
a small, shallow tray of some kind (you might like to make a simple one from cardboard or adapt the lid of a shoebox), marbles of 5 different colors.*

EVERYTHING around you is made up of tiny particles called atoms. Different things are made up of different atoms that are arranged in different ways. Crystals consist of atoms arranged in a regular, repeating pattern. This regular pattern inside a crystal gives it its consistent outer shape. This is why, no matter what their size, most crystals of a particular substance have exactly the same shape. Crystals form from gases, liquids or solutions that are cooling slowly. They increase in size when more and more atoms from the cooling gas or liquid add themselves to the surface of the crystal. In the first project on these pages, you can make a model of a crystal and see how the atoms build up inside it. Note that the model crystal you make is only a square pyramid because you use a square tray. Also, crystals grow in all directions, not just upward. The second project shows you how to grow your own real crystal very simply, from a liquid solution.

Make a model crystal

3 Add two more layers of marbles to make up a complete model crystal.

1 Fit a layer of marbles into the tray in a square pattern. Each atom is close to 8 others. In some substances, atoms are arranged in a hexagon (six-sided shape), with each one touching 6 others.

2 Add a second and a third layer. Each marble sits in a dip between 4 marbles in the layer below. Each marble in the third layer is directly above a marble in the very first layer.

Grow your own crystals

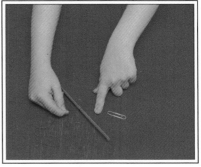

1 Get an adult to pour about 1 cup of very hot water into a pitcher. Add a spoonful of dish washing powder soap. Stir until it all dissolves. Add more soap until no more will dissolve.

2 Dissolving a solid in a liquid makes a solution. Your solution is said to be saturated because no more solid will dissolve. Pour the solution into a bowl; leave undissolved solids in the pitcher.

3 A crystal needs somewhere to start growing. Use a piece of thread to attach the paper clip to the straw. The distance from straw to clip should be about two-thirds the depth of the bowl.

M A T E R I A L S

You will need: pitcher, hot water, dish washing powder soap (sodium carbonate), spoon, bowl, straw, thread, paper clip, magnifying glass.

WARNING:
Do not rub your eyes when touching chemicals—wash your hands after each step.

4 As time goes by, some water evaporates, so there is not enough to keep all the solid dissolved. Some appears around the paper clip, and a clump of crystals starts to grow.

5 After several days, there will be a large clump of crystals growing on the paper clip. Remove the clip and crystals from the solution and wash them quickly under cold water. Now look closely at a crystal through a magnifying glass. You should be able to see that the shapes of your crystals are all the same.

STAIRWAYS TO HEAVEN

THE pyramids of ancient Egypt reached into the sky as stairways for the souls of their dead kings. The pyramid is the perfect shape to focus our gaze on the sky above. But, throughout history, other peoples have built differently shaped buildings with the same purpose. At the same time as the ancient Egyptians, the Chaldean people were building towers called ziggurats. The Chaldeans lived in the lands that are now Iran and Iraq, about 600 miles northeast of Egypt. Ziggurats were like stepped pyramids. They reached up into the sky so that Chaldean priests could be closer to their gods. All over the world, throughout history, different peoples have reached toward the skies. They do this because they believe that their gods live in heaven and that people's souls go to heaven when they die. Here are some differently shaped examples from around the world.

A caller summons people to prayer from the top of minarets

Great ziggurat, Ur, now Iraq
This platform is the lowest part of the ziggurat at Ur. It has been restored. Most ancient ziggurats were built of brick and crumbled long ago.

Minaret, Samarra, Iraq
This spiral minaret is 170 feet high. A minaret is a type of tower. You will find minarets in countries where Islam is the main religion. A person called a muezzin calls out from their tops several times a day to tell the faithful that it is time to pray. This minaret has a spiral ramp winding around the outside.

The Tower of Babel

This picture was painted in 1563, by Pieter Breughel. The tower is part of an ancient story described in the Bible. It tells how people tried to build a tower near Babylon to reach heaven. God stopped them by making each person speak a different language, so they could not understand each other. This story became one explanation of how so many different languages came about.

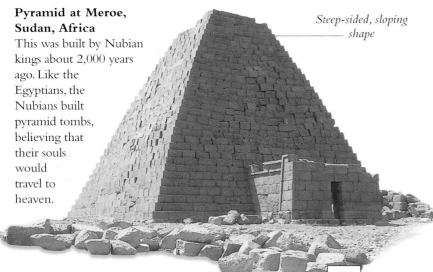

Steep-sided, sloping shape

Pyramid at Meroe, Sudan, Africa

This was built by Nubian kings about 2,000 years ago. Like the Egyptians, the Nubians built pyramid tombs, believing that their souls would travel to heaven.

Brihadeshwara temple, India

These temples are at Thanjavur, in southern India. Each pyramid-shaped roof is covered with intricate carvings. They show characters and ideas from Hindu legends.

MYTHS AND MYSTERIES

PYRAMIDS date from the earliest times of human civilization. Many things about them are mysterious, and there are many questions we cannot answer. Over the centuries and right up to the present, amazing stories have appeared that try to explain these mysteries. For example, how did the ancient Egyptians, Mayans and Aztecs move such enormous stones to build their pyramids? Some people say that visitors from outer space helped to move the stones. Look at the sarcophagus lid from the tomb at Palenque (page 41). One person suggested that it shows a spaceman taking off in a rocket. There are hundreds of other stories of this kind. However far-fetched the myths about pyramids, this simple shape has become a powerful symbol to everybody. It represents age, wisdom and something solid and truly long-lasting. But, more than anything else, the shape represents mystery—mysteries that we shall never completely understand.

In 1970, Thor Heyerdal sailed this copy of an ancient Egyptian boat from Africa to Central America. Some people think this proves that Egyptians crossed the Atlantic and taught the Mayans to build pyramids. However, people in different regions often come up with the same idea, and pyramids of the Americas appeared 2,000 years later.

Pyramid tomb, 1798–1806
This pyramid was attached to the tomb of Austrian Archduchess Marie Christine. In the late 1700s and early 1800s pyramid-shaped tombs were fashionable in Europe. Perhaps this was partly because the Egyptian pyramid stands for the idea of life continuing after death.

UXORI · OPTIMAE
ALBERTVS

PYRAMID MYTHS

• A blunt knife becomes sharp if it is placed under a pyramid!

• Pyramid-shaped hats improve a person's thinking power!

• Imhotep, the man who designed Djoser's step pyramid at Sakkara, Egypt, came from Antarctica!